Day Animals, Night Animals

Dona Herweck Rice

Publishing Credits

Rachelle Cracchiolo, M.S.Ed., *Publisher*
Conni Medina, M.A.Ed., *Managing Editor*
Nika Fabienke, Ed.D., *Content Director*
Véronique Bos, *Creative Director*
Shaun N. Bernadou, *Art Director*
Seth Rogers, *Editor*
John Leach, *Assistant Editor*
Courtney Roberson, *Senior Graphic Designer*

Image Credits: All images from iStock and/or Shutterstock.

Library of Congress Cataloging-in-Publication Data

Names: Rice, Dona, author.
Title: Day animals, night animals / Dona Herweck Rice.
Description: Huntington Beach, CA : Teacher Created Materials, [2019] | Audience: K to Grade 3. |
Identifiers: LCCN 2018029713 (print) | LCCN 2018031226 (ebook) | ISBN 9781493899531 | ISBN 9781493898794
Subjects: LCSH: Animal behavior--Juvenile literature. | Nocturnal animals--Juvenile literature.
Classification: LCC QL751.5 (ebook) | LCC QL751.5 .R544 2019 (print) | DDC 591.5--dc23
LC record available at https://lccn.loc.gov/2018029713

Teacher Created Materials
5301 Oceanus Drive
Huntington Beach, CA 92649-1030
www.tcmpub.com

ISBN 978-1-4938-9879-4

 call out in

Dogs

the  .

day

They call out to

each other.

 call out in

Owls

the  .

night

They call out to

each other.

 call out in

Cows

the  .

day

They call out to

each other.

 call out in

Frogs

the  .

night

They call out to

each other.

 call out in

Roosters

the  .

day

They call out to

each other.

High-Frequency Words

New Words

call
each
other
out

Review Words

in
the
they
to